REDBACK publishing

AF583934

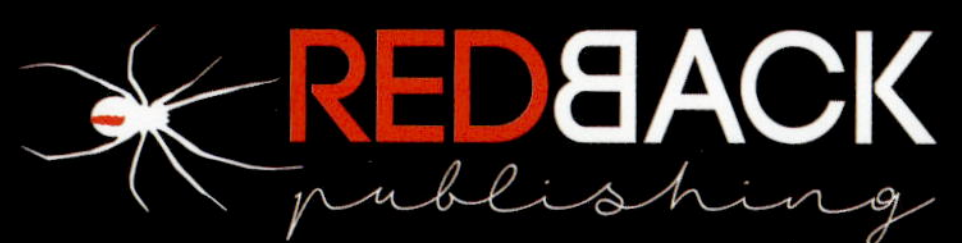

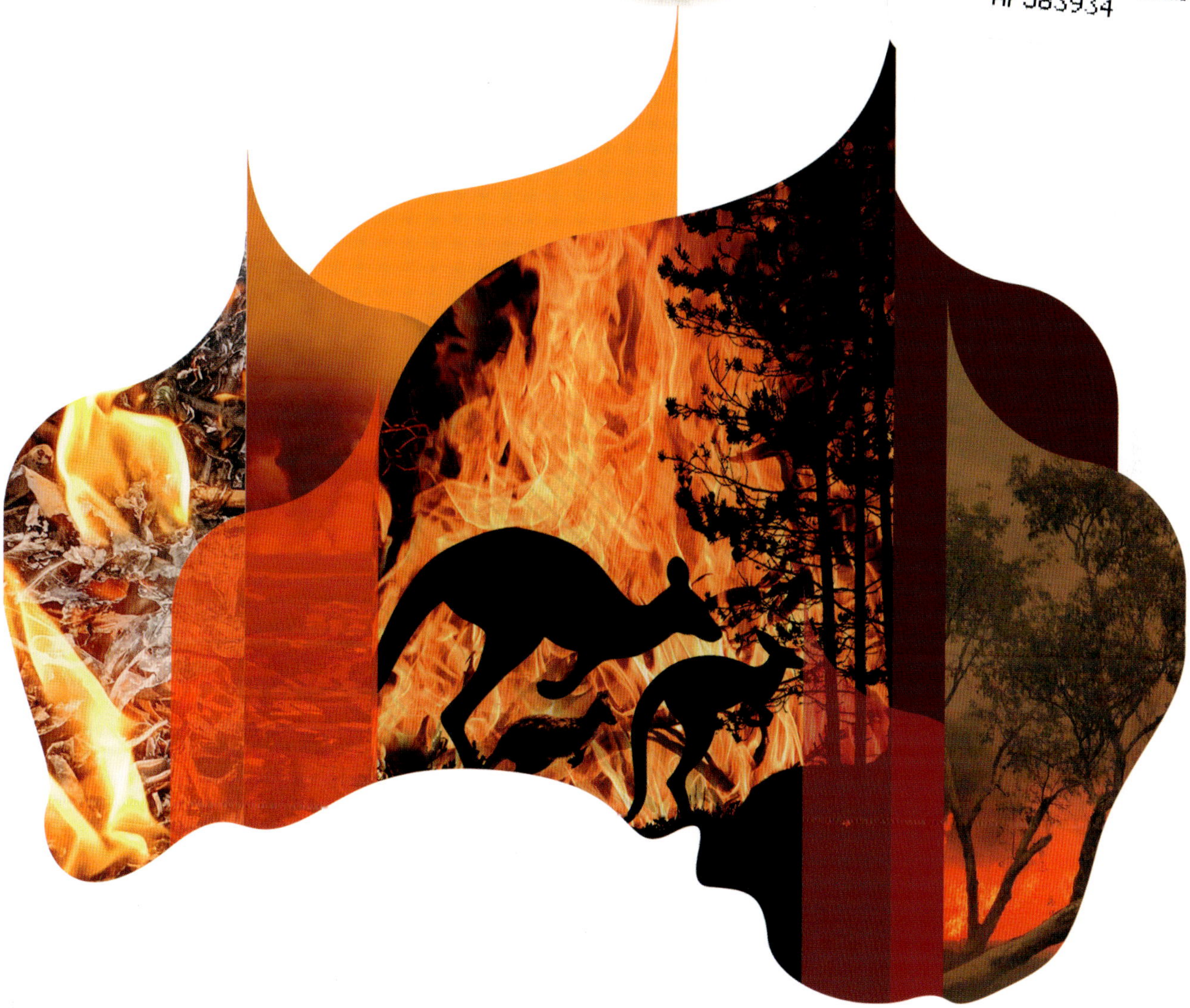

BUSHFIRES IN AUSTRALIA

BY JOHN LESLEY

Redback Publishing
PO Box 357 Frenchs Forest NSW 2086
Australia

www.redbackpublishing.com
orders@redbackpublishing.com

ISBN 978-1-922322-53-1

Author: John Lesley
Editor: Marlene Vaughan
Designer: Redback Publishing

Original illustrations © Redback Publishing 2021
Originated by Redback Publishing

Printed and bound in China

Acknowledgements
Abbreviations: l—left, r—right, b—bottom, t—top, c—centre, m—middle
We would like to thank the following for permission to reproduce photographs: (Images © shutterstock)
p4tr by JustLuna, p5bl by Daniel Charron, p9mr by Nils Versemann, p10l Museums Victoria, p10bm Tasmanian Archives and Heritage Office, p10r Sydney Oats / CC BY (https://creativecommons.org/licenses/by/2.0), p11br by Joachim Zens, p13br by Patrick Cooper, p20-21t by Petar B photography, p26t by Gayan DSA, p26m by ArliftAtoz2205, p26b by Claudine Van Massenhove

Every effort has been made to contact copyright holders of any material reproduced in this book. Any omissions will be rectified in subsequent printings if notice is given to the publisher.

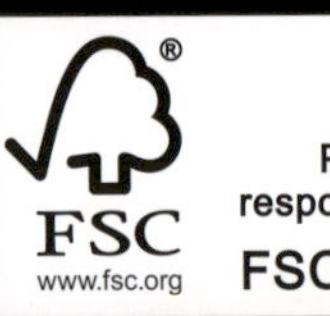

FSC
www.fsc.org
MIX
Paper from responsible sources
FSC® C020056

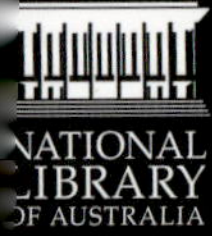

NATIONAL LIBRARY OF AUSTRALIA
A catalogue record for this book is available from the National Library of Australia

CONTENTS

BUSHFIRES IN AUSTRALIA

Along with cyclones and floods, bushfires have always been part of the Australian natural environment.

BUSHFIRES NEED THESE FOUR CONDITIONS BEFORE THEY CAN START AND SPREAD:

1. There needs to be enough vegetation to burn

2. The vegetation must be dry

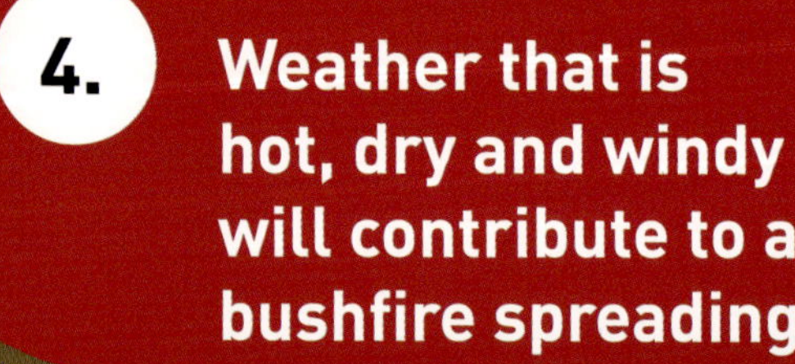

3. Something or someone starts the bushfire

4. Weather that is hot, dry and windy will contribute to a bushfire spreading

Increasing population has meant that more homes are being built in the outer parts of cities. These regions are close to natural bushland where there might be bushfires in the future.
NEWS
THE BUSHFIRES OF 2019/2020
Massive bushfires occurred in southeastern Australia at the end of 2019 and the beginning of 2020. Some of these fires destroyed everything in their path.

HOW DO BUSHFIRES START?

THE MAIN WAYS THAT BUSHFIRES START ARE:

1. Lightning strikes

2. Accidental fires started by people

3. Arson

Arsonists start bushfires on purpose. Arson is a criminal act. It is foolish, harms people and animals and it destroys beautiful places.

CLIMATE CHANGE

The Australian climate is getting warmer and drier. When the bush is dry, it burns easily and bushfires spread.

Although bushfires have been a natural hazard in Australia for many thousands of years, recent changes in climate seem to be having an impact on the size of bushfires we are experiencing today.

CONTROLLED BUSHFIRES

INDIGENOUS KNOWLEDGE

Australian Aboriginal people developed a system of controlled burning of the bush thousands of years ago.

These slow bushfires were carefully managed so that they removed only the undergrowth, and did not destroy whole forests.

In this way the landscape and the wildlife were kept safe from total destruction by fire.

FIREFIGHTERS AND CONTROLLED BURNING

Fire brigades across Australia undertake controlled burning of the bush to stop massive bushfires occurring later.

They pick days when there is not too much wind and the weather is not very hot. They also choose locations to start the burning so that the fire does not get out of control.

Aftermath of the 1983 Ash Wednesday bushfires at Mount Macedon

Residents survey what was left of the Victorian town of Omeo after the fire swept through.

Gore Street Mill on fire in South Hobart.

MASSIVE BUSHFIRES

Ruins of house burnt on Black Saturday bushfires in Victoria Australia

Dense bushfire smoke envelopes the Snowy Mountain Highway as people evacuate the South East Coast of New South Wales.

1983
ASH WEDNESDAY
in Victoria and South Australia

2009
BLACK SATURDAY
in Victoria

2019 / 2020
Bushfires raged for months across parts of southeastern Australia

PEOPLE IN BUSHFIRES

PEOPLE WHO LIVE IN AREAS THAT ARE CLOSE TO BUSHLAND SHOULD HAVE A BUSHFIRE SURVIVAL PLAN

Work out what you will do and where you will go if a bushfire is near your home in the future.

When will you leave to go to an evacuation centre or another safe place?

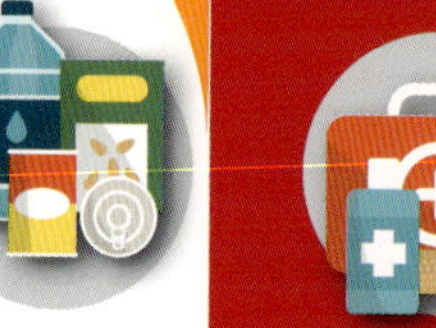

What will you take with you?

STAY OR LEAVE?

Every person who lives in a bushfire prone area should have a plan about when to leave if conditions are dangerous.

Local fire authorities have advice on how to decide what to do. It can be difficult to choose to leave your home, but your safety is more important than anything else.

Cars from evacuees parked on an oval during a bush fire crisis.

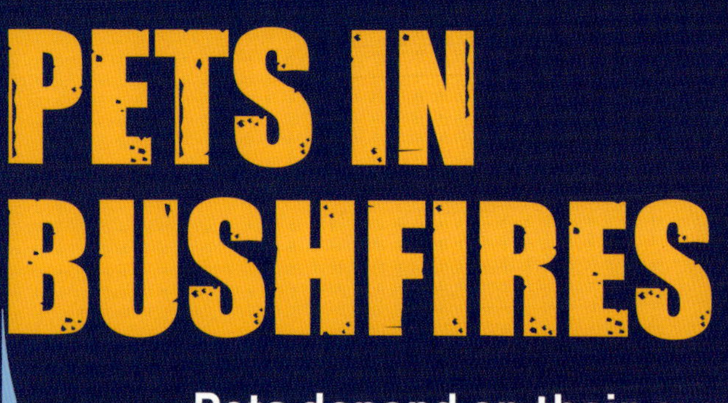

PETS IN BUSHFIRES

Pets depend on their owners so we need to have a plan about how we will keep them safe if there is a bushfire.

You could take them to stay with other people for a while, or take them to evacuation centres set up by the authorities when a bushfire threatens an area.

FARM ANIMALS IN BUSHFIRES

Cattle and sheep will suffer if a bushfire reaches grass in their paddocks.

Stock animals can sometimes be moved to safer areas before a bushfire reaches them. Choosing to do this should be part of a farmer's Bushfire Survival Plan.

WILD ANIMALS IN BUSHFIRES

Although there are many animal deaths in bushfires, some wild Australian animals have ways to survive.

Kangaroos will try to outrun a bushfire.

Koalas are very slow moving animals and cannot outrun a bushfire. Even if they do somehow survive, they will have no food if all the eucalyptus leaves were destroyed in the bushfire.

Wombats burrow deep under the ground.

Lizards also may hide under the ground.

FERAL ANIMALS IN BUSHFIRES

Feral pigs, deer and goats can run quickly to escape the flames. They move back to the burnt out areas to start eating all the new green shoots that start to grow after just a few days. This means there is less food for any native animal survivors of the bushfire.

Feral cats and dogs are clever and can move quickly. This helps them survive a bushfire. After the fire, feral carnivores hunt the native animals that have lost the safety of undergrowth or trees to hide in.

BUILDING IN BUSHFIRE AREAS

Governments have rules and regulations about ways that new buildings can be built in bushfire prone areas.

Sealing roof cavities so that embers cannot get in

Using extra tough glass in windows

Using plants in the garden that do not burn easily

Fitting shutters to outside doors and windows

EUCALYPTUS FORESTS

Eucalyptus trees have oils in their leaves. Once a fire reaches one of these trees the oil burns and sometimes causes the whole tree to burst into flames.

AFTER THE BUSHFIRE

Wildlife needs to be helped to find food and places to live

Local businesses that have lost buildings and stock will need help to start again

People whose homes have burnt down need help to rebuild or move somewhere else.

Bushfires are frightening, terrible events. After they have passed there is a lot of work to be done in the process of recovery.

People who have lost family, friends or pets in the bushfires will have feelings of grief for a long time

Rain can wash ash left by the fires into rivers, dams and creeks

Smoke in the air can continue to make breathing difficult, especially for people with asthma

GROWING BACK

The Australian bush has been subject to burning for so many thousands of years that many trees and bushes have developed special adaptations to survive.

Banksias have very hard seedpods that can survive fires and sprout once the fire is gone

Some wattles have seedpods that need a fire to make them open and start growing

Native grass trees will grow new green grass from their burnt stem

Some eucalyptus trees have hard buds on the trunk or just under the ground. These buds can often survive a bushfire and be ready to sprout new leaves very quickly afterwards

The ash from burnt plants is full of nutrients. Seeds can use this to start growing again.

FIREFIGHTERS

Bushfire fighting in Australia is the responsibility of teams. Their members undergo training and use special clothing, vehicles and equipment.

Firefighters from Australia have travelled to other countries to help fight fires there. In response, firefighters from overseas have been welcomed in Australia to help deal with local bushfires.

Thousands of years ago, Indigenous Australians developed ways of managing bushfires so that the landscapes they cared for would not suffer catastrophic destruction. Today, bushfire fighters continue to undertake hazard reduction burns to keep areas safe from monster bushfires.

VOLUNTEER BUSHFIRE BRIGADES Australians have enormous respect for the volunteer firefighters who place themselves in harm's way to help others.

FIRE TRUCKS

There are many different types of vehicles used to fight bushfires

Air Tankers - these control a bushfire by dropping water onto it

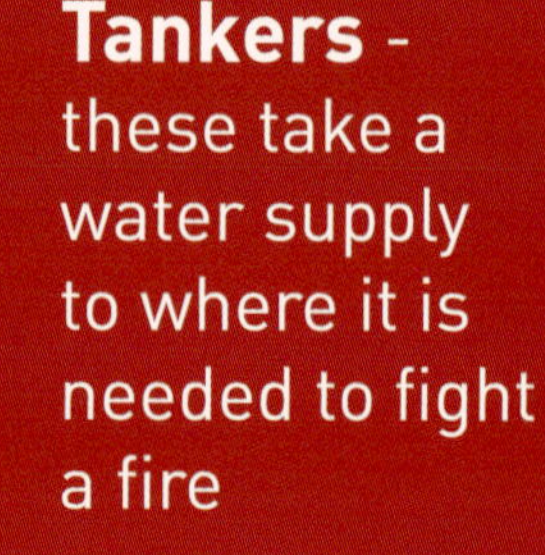

Tankers - these take a water supply to where it is needed to fight a fire

Pumpers - these use a local water source, even swimming pools, to fight fires

FIREFIGHTING GEAR

Firefighters need special clothing and equipment

Helmets

Jacket & Pants

Breathing masks

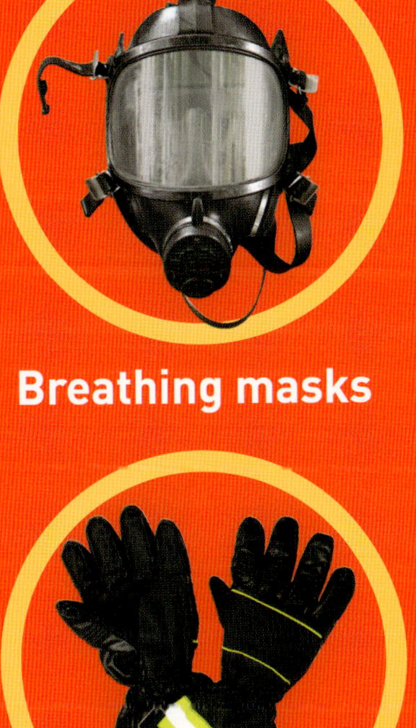

Gloves

Oxygen Tank

Boots

NEWS SERVICES

News services let us know what is happening when there is a bushfire in Australia.

Journalists have access to restricted areas so they can report on what is happening during a bushfire.

News reporters and their crews need to wear protective clothing when they work in dangerous conditions.

Journalists report for television, radio, online news services and newspapers. We do not see the camera operators and technical crew, but they work in the bushfire area as well.

FIRE DANGER RATINGS

On days when there is high bushfire risk, emergency services show the Fire Danger Rating on signs posted in public areas.

Local agencies can declare a Total Fire Ban if conditions are dangerous. In a Total Fire Ban, people cannot light any sort of fire outdoors, not even a barbeque if it uses wood or heat beads.

WORDS ABOUT BUSHFIRES

Outside Australia people have different names for fires that occur in natural areas:

WILDFIRES
BRUSH FIRES
FOREST FIRES

arsonist criminal who starts a bushfire on purpose

asthma illness that makes breathing difficult

carnivore animal that eats meat

cyclone storm with rain and very strong winds

ember little piece of burning wood or other material

evacuate leave a place which is dangerous

feral animal animal that is not native to an area

fuel load amount of vegetation that could burn in a fire

Hazard Reduction Burn controlled burning of bushland

journalist person who reports on news events

massive extremely large

restricted place place where only some people are allowed to be

roof cavity space between a ceiling and the roof

spotting fires started by burning embers blown by the wind

suburban in a district near a city

undergrowth plants growing under trees

vegetation plants and trees

volunteer person who works without pay to help others

INDEX